Asterisk Cookbook

Asterisk Cookbook

Leif Madsen and Russell Bryant

O'REILLY®

Beijing · Cambridge · Farnham · Köln · Sebastopol · Tokyo

Asterisk Cookbook

by Leif Madsen and Russell Bryant

Published by O'Reilly Media, Inc., 1005 Gravenstein Highway North, Sebastopol, CA 95472.

O'Reilly books may be purchased for educational, business, or sales promotional use. Online editions are also available for most titles (*http://my.safaribooksonline.com*). For more information, contact our corporate/institutional sales department: (800) 998-9938 or *corporate@oreilly.com*.

Editor:	Mike Loukides	**Cover Designer:**	Karen Montgomery
Production Editor:	Adam Zaremba	**Interior Designer:**	David Futato
Proofreader:	Adam Zaremba	**Illustrator:**	Robert Romano

Printing History:

April 2011:	First Edition.

ISBN: 978-1-449-30382-2

[LSI]

1301331848

Table of Contents

Preface

This is a book for anyone who uses Asterisk, but particularly those who already understand the dialplan syntax.

In this book, we look at common problems we've encountered as Asterisk administrators and implementers, then show solutions to those problems using the Asterisk dialplan. As you go through the recipes and start looking at the solutions, you may think, "Oh, that's a neat idea, but they could have also done it this way." That might happen a lot, because with Asterisk, the number of solutions available for a particular problem are astounding. We have chosen to focus on using the tools available to us within Asterisk, so solutions heavily focus on the use of dialplan, but that doesn't mean an external application through the Asterisk Gateway Interface or Asterisk Manager Interface isn't also possible.

Readers of this book should be familiar with many core concepts of Asterisk, which is why we recommend that you already be familiar with the content of *Asterisk: The Definitive Guide* (*http://oreilly.com/catalog/9780596517342/*), also published by O'Reilly. This book is designed to be a complement to it.

We hope you find some interesting solutions in this book that help you to be creative in future problem solving.

Organization

The book is organized into these chapters:

Chapter 1, *Dialplan Fundamentals*
 This chapter shows some examples of fundamental dialplan constructs that will be useful over and over again.

Chapter 2, *Call Control*
 This chapter discusses a number of examples of controlling phone calls in Asterisk.

Chapter 3, *Audio Manipulation*
 This chapter has examples of ways to get involved with the audio of a phone call.

Software

This book is focused on documenting Asterisk Version 1.8; however, many of the conventions and information in this book are version-agnostic.

Conventions Used in This Book

The following typographical conventions are used in this book:

Italic

> Indicates new terms, URLs, email addresses, filenames, file extensions, pathnames, directories, and Unix utilities.

`Constant width`

> Indicates commands, options, parameters, and arguments that must be substituted into commands.

`Constant width bold`

> Shows commands or other text that should be typed literally by the user. Also used for emphasis in code.

`Constant width italic`

> Shows text that should be replaced with user-supplied values.

`[ Keywords and other stuff ]`

> Indicates optional keywords and arguments.

`{ choice-1 | choice-2 }`

> Signifies either *choice-1* or *choice-2*.

 This icon signifies a tip, suggestion, or general note.

 This icon indicates a warning or caution.

Using Code Examples

This book is here to help you get your job done. In general, you may use the code in this book in your programs and documentation. You do not need to contact us for permission unless you're reproducing a significant portion of the code. For example, writing a program that uses several chunks of code from this book does not require permission. Selling or distributing a CD-ROM of examples from O'Reilly books does require permission. Answering a question by citing this book and quoting example code does not require permission. Incorporating a significant amount of example code from this book into your product's documentation does require permission.

We appreciate, but do not require, attribution. An attribution usually includes the title, author, publisher, and ISBN. For example: "*Asterisk Cookbook*, First Edition, by Leif Madsen and Russell Bryant (O'Reilly). Copyright 2011 Leif Madsen and Russell Bryant, 978-1-449-30382-2."

If you feel your use of code examples falls outside fair use or the permission given above, feel free to contact us at *permissions@oreilly.com*.

Safari® Books Online

 Safari Books Online is an on-demand digital library that lets you easily search over 7,500 technology and creative reference books and videos to find the answers you need quickly.

With a subscription, you can read any page and watch any video from our library online. Read books on your cell phone and mobile devices. Access new titles before they are available for print, and get exclusive access to manuscripts in development and post feedback for the authors. Copy and paste code samples, organize your favorites, download chapters, bookmark key sections, create notes, print out pages, and benefit from tons of other time-saving features.

O'Reilly Media has uploaded this book to the Safari Books Online service. To have full digital access to this book and others on similar topics from O'Reilly and other publishers, sign up for free at *http://my.safaribooksonline.com*.

How to Contact Us

Please address comments and questions concerning this book to the publisher:

O'Reilly Media, Inc.
1005 Gravenstein Highway North
Sebastopol, CA 95472
800-998-9938 (in the United States or Canada)
707-829-0515 (international or local)
707-829-0104 (fax)

We have a web page for this book, where we list errata, examples, and any additional information. You can access this page at:

http://oreilly.com/catalog/9781449303822/

To comment or ask technical questions about this book, send email to:

bookquestions@oreilly.com

For more information about our books, courses, conferences, and news, see our website at *http://www.oreilly.com.*

Find us on Facebook: *http://facebook.com/oreilly*

Follow us on Twitter: *http://twitter.com/oreillymedia*

Watch us on YouTube: *http://www.youtube.com/oreillymedia*

Acknowledgments

While writing this book, we used O'Reilly's Open Feedback Publishing System (OFPS), which allowed Asterisk community members to read and comment on the content as we were writing it. The following people provided feedback to us via OFPS: Stefan Schmidt, Scott Howell, Christian Gutierrez, Jason "not a nub" Parker, and Paul Belanger. Thank you all for your valuable contributions to this book!

CHAPTER 1

Dialplan Fundamentals

1.0 Introduction

This chapter is designed to show you some fundamental dialplan usage concepts that we use in nearly every dialplan. We've developed these recipes to show you how we've found the usage of these dialplan applications to be of the greatest value and flexibility.

1.1 Counting and Conditionals

Problem

You need to perform basic math—such as increasing a counting variable—and do it using a conditional statement.

Solution

In many cases you will need to perform basic math, such as when incrementing the counter variable when performing loops. To increase the counter variable we have a couple of methods which are common. First, we can use the standard conditional matching format in Asterisk:

```
[CounterIncrement]
exten => start,1,Verbose(2,Increment the counter variable)

; Set the initial value of the variable
    same => n,Set(CounterVariable=1)
    same => n,Verbose(2,Current value of CounterVariable is: ${CounterVariable})

; Now we can increment the value of CounterVariable
    same => n,Set(CounterVariable=$[${CounterVariable} + 1])
    same => n,Verbose(2,Our new value of CounterVariable is: ${CounterVariable})

    same => n,Hangup()
```

Alternatively, in versions of Asterisk greater than and including Asterisk 1.8, we can use the INC() dialplan function:

```
[CounterIncrement]
exten => start,1,Verbose(2,Increment the counter variable)

; Set the inital value of the variable
    same => n,Set(CounterVariable=1)
    same => n,Verbose(2,Current value of CounterVariable is: ${CounterVariable})

; Now we can increment the value of CounterVariable
    same => n,Set(CounterVariable=${INC(CounterVariable)})
    same => n,Verbose(2,Our new value of CounterVariable is: ${CounterVariable})

    same => n,Hangup()
```

Additionally, we can use the IF() function to determine whether we should be incrementing the value at all:

```
[CounterIncrement]
exten => start,1,Verbose(2,Increment the counter variable)

; Set the inital value of the variable
    same => n,Set(CounterVariable=1)
    same => n,Verbose(2,Current value of CounterVariable is: ${CounterVariable})

; Here we use the RAND() to randomly help us determine whether we should increment
; the CounterVariable.  We've set a range of 0 through 1, which we'll use
; as false (0) or true (1)
    same => n,Set(IncrementValue=${RAND(0,1)})

; Now we can increment the value of CounterVariable if IncrementValue returns 1
    same => n,Set(CounterVariable=${IF($[${IncrementValue} = 1]?
${INC(CounterVariable)}:${CounterVariable})
    same => n,Verbose(2,Our IncrementValue returned: ${IncrementValue})
    same => n,Verbose(2,Our new value of CounterVariable is: ${CounterVariable})

    same => n,Hangup()
```

Discussion

The incrementing of variables in Asterisk is one of the more common functionalities you'll encounter, especially as you start building more complex dialplans where you need to iterate over several values. In versions of Asterisk prior to 1.8, the most common method for incrementing (and decrementing) the value of a counter variable was with the use of the dialplan conditional matching format, which we explored in the first example.

With newer versions of Asterisk, the incrementing and decrementing of variables can be done using the INC() and DEC() dialplan functions respectively. The use of the INC() and DEC() functions requires you to specify only the name of the variable you want to change, and a value is then returned. You do not specify the value to change, which means you would provide ${INC(CounterVariable)}, not ${INC(${CounterVaria ble})}. If the value of ${CounterVariable} returned 5, then INC() would try and increment the value of the channel variable 5, and not increment the value of 5 to the value 6.

 Of course, you could pass ${MyVariableName} to the INC() function, and if the value of ${MyVariableName} contained the name of the variable you actually wanted to retrieve the value of, increment, and then return, you could do that.

It should also be explicitly stated that INC() and DEC() do not modify the original value of the variable. They simply check to see what the current value of the channel variable is (e.g., 5), increment the value in memory, and return the new value (e.g., 6).

In our last code block, we've used the RAND() dialplan function to return a random 0 or 1 and assign it to the IncrementVariable channel variable. We then use the value stored in ${IncrementVariable} to allow the IF() dialplan function to return either an incremented value or the original value. The IF() function has the syntax:

```
${IF($[...conditional statement...]?true_return_value:false_return_value)}
```

The true or false return values are optional (although you need to return something in one of them). In our case we set our *true_return_value* to ${INC(CounterVariable)}, which would return an incremented value that would then be assigned to the Counter Variable channel variable. We set the *false_return_value* to ${CounterVariable}, which would return the original value of ${CounterVariable} to the CounterVariable channel variable, thereby not changing the value.

What we've described are common methods for incrementing (and decrementing) channel variables in the dialplan that we've found useful.

See Also

Recipe 1.2

1.2 Looping in the Dialplan

Problem

You need to perform an action several times before continuing on in the dialplan.

Solution

A basic loop which iterates a certain number of times using a counter can be created in the following manner:

```
[IteratingLoop]
exten => start,1,Verbose(2,Looping through an action five times.)
    same => n,Set(X=1)
    same => n,Verbose(2,Starting the loop)
    same => n,While($[${X} <= 5])
    same => n,Verbose(2,Current value of X is: ${X})
    same => n,Set(X=${INC(X)})
    same => n,EndWhile()
    same => n,Verbose(2,End of the loop)
    same => n,Hangup()
```

We could build the same type of counter-based loop using the GotoIf() application as well:

```
[IteratingLoop]
exten => start,1,Verbose(2,Looping through an action five times.)
    same => n,Set(X=1)
    same => n,Verbose(2,Starting the loop)
    same => n(top),NoOp()
    same => n,Verbose(2,Current value of X is: ${X})
    same => n,Set(X=${INC(X)})
    same => n,GotoIf($[${X} <= 5]?top)
    same => n,Verbose(2,End of the loop)
    same => n,Hangup()
```

Sometimes you might have multiple values you want to check for. A common way of iterating through several values is by saving them to a variable and separating them with a hyphen. We can then use the CUT() function to select the field that we want to check against:

```
[LoopWithCut]
exten => start,1,Verbose(2,Example of a loop using the CUT function.)
    same => n,Set(Fruits=Apple-Orange-Banana-Pineapple-Grapes)
    same => n,Set(FruitWeWant=Pineapple)
    same => n,Set(X=1)
    same => n,Set(thisFruit=${CUT(Fruits,-,${X})})
    same => n,While($[${EXISTS(${thisFruit})}])
    same => n,GotoIf($[${thisFruit} = ${FruitWeWant}]?GotIt,1)
    same => n,Set(X=${INC(X)})
    same => n,thisFruit=${CUT(Fruits,-,${X})})
    same => n,EndWhile()

; We got to the end of the loop without finding what we were looking for.
    same => n,Verbose(2,Exiting the loop without finding our fruit.)
    same => n,Hangup()

; If we found the fruit, then the GotoIf() will get us here.
exten => GotIt,1,Verbose(2,We matched the fruit we were looking for.)
    same => n,Hangup()
```

Discussion

We've explored three different blocks of code which show common ways of performing loops in the Asterisk dialplan. We've shown two ways of performing a counting-based loop, which lets you iterate through a set of actions a specified number of times. The first method uses the While() and EndWhile() applications to specify the bounds of the loop, with the check happening at the top of the loop. The second method uses the GotoIf() application to check whether the loop continues at the bottom of the loop block.

The third loop we've shown uses the CUT() dialplan function to move through fields in a list of words that we check for in our loop. When we find what we're looking for, we jump to another location in the dialplan (the GotIt extension), where we can then continue performing actions knowing we've found what we're looking for. If we iterate through the loop enough times, the channel variable thisFruit will contain nothing, and the loop will then continue at the EndWhile() application, falling through to the rest of the priorities below it. If we get there, we know we've fallen out of our loop without finding what we're looking for.

There are other variations on these loops, such as with the use of the Continue While() and ExitWhile() applications, and the method with which we search for data can also be different, such as with the use of the ARRAY() and HASH() dialplan functions, which are useful when returning data from a relational database using *func_odbc*.

See Also

Recipe 1.1

1.3 Controlling Calls Based on Date and Time

Problem

When receiving calls in your auto-attendant, you sometimes need to redirect calls to a different location of the dialplan based on the date and/or time of day.

Solution

```
[AutoAttendant]
exten => start,1,Verbose(2,Entering our auto-attedant)
    same => n,Answer()
    same => n,Playback(silence/1)

; We're closed on New Years Eve, New Years Day, Christmas Eve, and Christmas Day
    same => n,GotoIfTime(*,*,31,dec?holiday,1)
    same => n,GotoIfTime(*,*,1,jan?holiday,1)
    same => n,GotoIfTime(*,*,24,dec?holiday,1)
    same => n,GotoIfTime(*,*,25,dec?holiday,1)
```

```
; Our operational hours are Monday-Friday, 9:00am to 5:00pm.
    same => n,GotoIfTime(0900-1700,mon-fri,*,*?open,1:closed,1)

exten => open,1,Verbose(2,We're open!)
    same => n,Background(custom/open-greeting)
...


exten => closed,1,Verbose(2,We're closed.)
    same => n,Playback(custom/closed-greeting)
    same => n,Voicemail(general-mailbox@default,u)
    same => n,Hangup()

exten => holiday,1,Verbose(2,We're closed for a holiday.)
    same => n,Playback(custom/closed-holiday)
    same => n,Voicemail(general-mailbox@default,u)
    same => n,Hangup()
```

We don't just need to use the GotoIfTime() application in an auto-attendant. Sometimes we want to forward calls to people based on time, such as when IT staff is not in the office on weekends, but are on call:

```
[Devices]
exten => 100,1,Verbose(2,Calling IT Staff.)
    same => n,GotoIfTime(*,sat&sun,*,*?on_call,1)
    same => n,Dial(SIP/itstaff,30)
    same => n,Voicemail(itstaff@default,u)
    same => n,Hangup()

exten => on_call,1,Verbose(2,Calling On-Call IT Staff.)
    same => n,Dial(SIP/myITSP/4165551212&SIP/myITSP/2565551212,30)
    same => n,Voicemail(itstaff@default,u)
    same => n,Hangup()
```

Discussion

At the top of our auto-attendant, we Answer() the call and Playback(silence/1) which are standard actions prior to playing back prompts to the caller. This eliminates the first few milliseconds of a prompt from being cut off. After that, we then start our checks with the GotoIfTime() application. We start with specific matches first (such as particular days of the week) before we do our more general checks, such as our 9:00 a.m. to 5:00 p.m., Monday to Friday checks. If we did it the other way around, then we'd be open Monday–Friday, 9:00 a.m.–5:00 p.m. on holidays.

The GotoIfTime() application contains five fields (one of which is optional; we'll discuss it momentarily). The fields are: time range, days of the week, days of the month, and months. The fifth field, which is optional and specified after the months field, is the timezone field. If you are servicing multiple timezones, you could use this to have different menus and groups of people answering the phones between 9:00 a.m. and 5:00 p.m. for each timezone.

On holidays, we have a separate part of the menu in which we play back a prompt and then send the caller to Voicemail(), but you could, of course, send the caller to any functionality that you wish. We've used the asterisk (*) symbol to indicate that at any time of the day and any day of the week, but only on the 31st day of December, should calls go to the holiday extension. We then specify in the same manner for the 1st of January, the 24th of December, and the 25th of December, all of which are handled by the holiday extension.

After we've determined it's not a holiday, then we perform our standard check to see if we should be using the open extension—which we'll use when the office is open—or the closed extension—which plays back a prompt indicating we're closed and to leave a message in the company Voicemail().

In our second block of code, we've also shown how you could use the GotoIfTime() to call IT staff on the weekends (Saturday and Sunday, any time of the day). Normally, people would call the extension 100, and if it is Monday to Friday, they would be directed to the SIP device registered to [itstaff] in *sip.conf*. Of course, since the people reading and implementing this probably are IT staff, there is a good chance you will alter this to call at hours which are more sane.

1.4 Authenticating Callers

Problem

You need to authenticate callers prior to moving on in the dialplan.

Solution

```
[Authentication]
exten => start,1,Verbose(2,Simple Authenicate application example)
    same => n,Playback(silence/1)
    same => n,Authenticate(1234)
    same => n,Hangup()
```

Here is a slightly modified version that sets the maxdigits value to 4, thereby not requiring the user to press the # key when done entering the password:

```
[Authentication]
exten => start,1,Verbose(2,Simple Authenicate application example)
    same => n,Playback(silence/1)
    same => n,Authenticate(1234,,4)
    same => n,Hangup()
```

By starting our password field with a leading forward slash (/), we can utilize an external file as the source of the password(s):

```
[Authentication]
exten => start,1,Verbose(2,Simple Authenicate application example)
    same => n,Playback(silence/1)
```

```
    same => n,Authenticate(/etc/asterisk/authenticate/passwd_list.txt)
    same => n,Hangup()
```

If we use the d flag, Asterisk will interpret the path provided as a database key in the Asterisk DB instead of a file:

```
[Authentication]
exten => start,1,Verbose(2,Simple Authenicate application example)
    same => n,Playback(silence/1)
    same => n,Authenticate(/authenticate/password,d)
    same => n,Hangup()
```

We can insert and modify the password we're going to use in the Asterisk database using the Asterisk CLI:

```
*CLI> database put authenticate/password 1234 this_does_not_matter
Updated database successfully

*CLI> database show authenticate
/authenticate/password/1234                  : this_does_not_matter
1 result found.
```

A neat modification for temporary passwords in the Asterisk Database (AstDB) is by adding the r flag along with the d flag to remove the password from the AstDB upon successful authentication:

```
[Authentication]
exten => start,1,Verbose(2,Simple Authenicate application example)
    same => n,Playback(silence/1)
    same => n,Authenticate(/authenticate/temp_password,dr)
    same => n,Hangup()
```

After we insert the password into the database, we can use it until an authentication happens, and then it is removed from the database:

```
*CLI> database put authenticate/temp_password 1234 this_does_not_matter
Updated database successfully

*CLI> database show authenticate
```

Discussion

The Authenticate() dialplan application is quite basic at its core; a password is provided to the dialplan application, and that password must be entered correctly to authenticate and continue on in the dialplan. The Authenticate() application is one of the older applications in Asterisk, and it shows that by the number of available options for something that should be an almost trivial application. While much of the functionality provided by the Authenticate() application can be done in the dialplan using dialplan functions and other applications, it is somewhat nice to have much of the functionality contained within the application directly.

We've provided examples of some of this functionality in the Solution section. We started off with a simple example of how you can provide a password to Authenti cate() and then require that password to be entered before continuing on in the dialplan. The first example required the user to enter a password of 1234 followed by the # key to signal that entering of digits was complete. Our second example shows the use of the *maxdigits* field, with a value of 4, to not require the user to press the # key when done entering the password.

We went on to show how you could provide a path to the password field, which would allow you to utilize a file for authentication. Using an external file can be useful if you want to use an external script to rotate the password fairly often. One of the particular uses we can think of would be the SecurID system, by RSA, which uses cards that contain a number that rotates over a period of time to be synchronized with a centralized server that is using the same algorithm to generate passwords. If you wanted to tie this system into Asterisk, you could have a script that rotated the key on a timely basis.

Instead of using a file for the location of the password, there is an option that lets you use the Asterisk DB. With the d flag, we can tell Authenticate() that the path we're providing is that of a family/key relationship in the AstDB. An additional flag, r, can also be provided that removes the key from the database upon successful authentication, which provides a method for one-time-use passwords.

The Authenticate() application is a general purpose authentication mechanism which provides a base layer without taking into consideration which user or caller is attempting to authenticate. It would be fairly straightforward, however, to add that layer with some additional dialplan, or you could even utilize some of the dialplan we've provided in this book.

See Also

Recipe 1.5, Recipe 1.6

1.5 Authenticating Callers Using Voicemail Credentials

Problem

You need to provide an authentication mechanism in your dialplan, but wish to use the credentials already in place for retrieving voicemail.

Solution

```
[Authentication]
exten => start,1,Verbose(2,Attempting to authenticate caller with voicemail creds.)
    same => n,Playback(silence/1)

; This is where we do our authentication
    same => n,VMAuthenticate(@default)

    same => n,Verbose(2,The caller was authenticated if we execute this line.)
    same => n,Goto(authenticated,1)
```

```
exten => authenticated,1,Verbose(2,Perform some actions.)
    ; would contain 'default'.
    same => n,Verbose(2,Value of AUTH_CONTEXT: ${AUTH_CONTEXT})
    ; mailbox '100' was authenticated.
    same => n,Verbose(2,Value of AUTH_MAILBOX: ${AUTH_MAILBOX})
    same => n,Playback(all-your-base)
    same => n,Hangup()
```

If you wanted to explicitly define the mailbox to authenticate against, you could place the extension number to authenticate with in front of the voicemail context:

```
same => n,VMAuthenticate(100@default)
```

And if you don't like the introductory prompts that VMAuthenticate() plays, you could modify your dialplan to play a different initial prompt by adding the s flag:

```
same => n,Playback(extension)
same => n,VMAuthenticate(@default,s)
```

Also, if we wanted to provide the option of letting someone skip out of authenticating altogether, and perhaps speak with an operator, we could provide the a extension in our context, which allows the user to press * to jump to the a extension:

```
[Authentication]
exten => start,1,Verbose(2,Attempting to authenticate caller with voicemail creds.)
    same => n,Playback(silence/1)

; This is where we do our authentication
    same => n,VMAuthenticate(@default)

    same => n,Verbose(2,The caller was authenticated if we execute this line.)
    same => n,Goto(authenticated,1)

exten => a,1,Verbose(2,Calling the operator.)
    same => n,Dial(SIP/operator,30)
    same => n,Voicemail(operator@default,u)
    same => n,Hangup()
```

Discussion

The VMAuthenticate() application provides us a useful tool for authenticating callers using credentials the user already knows, and uses often. By using the caller's own voicemail box number and password, it is one less piece of authentication information to be memorized on the part of the caller. It also provides a centralized repository of credentials the administrator of the system can control and enforce to be secure. Since we have the ability to skip playing back the initial prompts, we could obtain the mailbox number used for authentication in several ways: by asking the user to provide it using another application or a lookup from a database, or simply by dialing it from the phone and performing a pattern match.

Another advantage is providing the ability to quit out of the authentication mechanism and be connected to a live agent, who can then perform the authentication for the caller using information stored in his database and then transfer the caller to the part of the system which would have required her authentication credentials.

See Also

Recipe 1.6, Recipe 1.4

1.6 Authenticating Callers Using Read()

Problem

You want to authenticate callers using a custom set of prompts and logic.

Solution

A basic system which uses a static pin number for authentication:

```
[Read_Authentication]
exten => start,1,NoOp()
    same => n,Playback(silence/1)
    same => n,Set(VerificationPin=1234)    ; set pin to verify against
    same => n,Set(TriesCounter=1)          ; counter for login attempts
    same => n(get_pin),Read(InputPin,enter-password)  ; get a pin from
                                                        ; from the caller

; Check if the pin input by the caller is the same as our verification pin.
    same => n,GotoIf($["${InputPin}" = "${VerificationPin}"]?pin_accepted,1)

; Increment the TriesCounter by 1
    same => n,Set(TriesCounter=${INC(TriesCounter)})
    same => n,GotoIf($[${TriesCounter} > 3]?too_many_tries,1)
    same => n,Playback(vm-incorrect)
    same => n,Goto(get_pin)

exten => pin_accepted,1,NoOp()
    same => n,Playback(auth-thankyou)
    same => n,Hangup()

exten => too_many_tries,1,NoOp()
    same => n,Playback(vm-incorrect)
    same => n,Playback(vm-goodbye)
    same => n,Hangup()
```

We can modify the dialplan slightly to provide greater security by requiring an account code and a pin, which can be loaded from the AstDB:

```
[Read_Authentication]
exten => start,1,NoOp()
    same => n,Playback(silence/1)

; Set a couple of counters for login attempts
    same => n,Set(TriesCounter=1)
    same => n,Set(InvalidAccountCounter=1)
```

```
; Request the access code (account code)
    same => n(get_acct),Read(InputAccountNumber,access-code)
; make sure we have an account number
    same => n,GotoIf($[${ISNULL(${InputAccountNumber})}]?get_acct)

; Request the password (pin)
    same => n(get_pin),Read(InputPin,vm-password)

; Check the database to see if a password exists for the account number entered
    same => n,GotoIf($[${DB_EXISTS(access_codes/${InputAccountNumber})}]?
have_account,1:no_account,1)

; If a pin number exists check it against what was entered
exten => have_account,1,NoOp()
    same => n,Set(VerificationPin=${DB_RESULT})
    same => n,GotoIf($["${InputPin}" = "${VerificationPin}"]?pin_accepted,1)
    same => n,Set(TriesCounter=${INC(TriesCounter)})
    same => n,GotoIf($[${TriesCounter} > 3]?too_many_tries,1)
    same => n,Playback(vm-incorrect)
    same => n,Goto(start,get_pin)

; If no account exists, request a new access code be entered
exten => no_account,1,NoOp()
    same => n,Playback(invalid)
    same => n,Set(InvalidAccountCounter=${INC(InvalidAccountCounter)})
    same => n,GotoIf($[${InvalidAccountCounter} > 3]?too_many_tries,1)
    same => n,Goto(start,get_acct)

; Account and pin were verified
exten => pin_accepted,1,NoOp()
    same => n,Playback(auth-thankyou)
    same => n,Hangup()

; Sorry, too many attempts to login. Hangup.
exten => too_many_tries,1,NoOp()
    same => n,Playback(vm-incorrect)
    same => n,Playback(vm-goodbye)
    same => n,Hangup()
```

We added the access code (555) and pin (1234) to the AstDB using the following command:

```
*CLI> database put access_codes 555 1234
```

Discussion

If you've already looked at the solutions in Recipes 1.5 and 1.4, what you'll immediately notice is that the other solutions are quite a bit more compact. The main reason is that all of the loop control is handled within the dialplan application, whereas in this case we're defining the loop control with dialplan logic and handling the number of loops and what to do on failure ourselves. With greater control comes greater complexity, and that is illustrated in the examples provided. The advantage to this greater level of verbosity is the control at each step: which prompts are played, when they're played, and how often they are played. We also get to control the method of authentication.

In our first example, we showed a fairly basic authentication method which simply asked the caller for a password that we statically defined within the dialplan.* After entering the pin, we check what the user entered against what we set in the Verifica tionPin channel variable. If the numbers do not match, we increment a counter, test to see the number has increased to greater than 3, and, if not, request that the caller re-enter her pin. If the number of tries exceeds 3, then we play a prompt saying goodbye and hang up.

Our second example was expanded to include both an access code (account code) and a password (pin) which we've written to the AstDB. When the caller enters the dialplan, we request an access code and a password. We then check the AstDB using DB_EXISTS() to determine if the account number exists in the database. If it does not, then we inform the user that the account does not exist using a dialplan defined by the no_account extension. This is followed by a check to determine if the caller has entered an invalid account number and, if so, determine if this has happened more than 3 times, in which case we then disconnect the caller.

If the caller has entered a valid account number, we then handle additional logic in the have_account extension, where we verify the pin number entered against what was returned from the database. If everything is valid, then we play back a prompt in the pin_accepted extension, and hang up the call (although it's implied additional logic could then be handled now that the caller has been validated).

Because the solutions described contain a lot of dialplan logic and aren't tied to any particular dialplan application (other than Read(), which we're using for data input), the solutions could easily be modified to authenticate against other external sources of data. For example, the REALTIME() functions could be used to gather credentials from an LDAP database, or *func_odbc* could be employed to gather information from a relational database. You could even use CURL() to pass the data collected from the caller to authenticate against a web page which could then return account information to the caller. The possibilities with custom dialplan really are only limited by the problems encountered and solved by your imagination.

See Also

Recipe 1.5, Recipe 1.4, Recipe 1.2, Recipe 1.1

---

* Of course we could have defined that as a global variable in the [globals] section.

1.7 Debugging the Dialplan with Verbose()

Problem

You would like to insert debugging into your dialplan that can be enabled on a global, per-device, or per-channel basis.

Solution

Use something like this `chanlog GoSub()` routine:

```
[chanlog]
exten => s,1,GotoIf($[${DB_EXISTS(ChanLog/all)} = 0]?checkchan1)
    same => n,GotoIf($[${ARG1} <= ${DB(ChanLog/all)}]?log)
    same => n(checkchan1),Set(KEY=ChanLog/channel/${CHANNEL})
    same => n,GotoIf($[${DB_EXISTS(${KEY})} = 0]?checkchan2)
    same => n,GotoIf($[${ARG1} <= ${DB(${KEY})}]?log)
    same => n(checkchan2),Set(KEY=ChanLog/channel/${CUT(CHANNEL,-,1)})
    same => n,GotoIf($[${DB_EXISTS(${KEY})} = 0]?return)
    same => n,GotoIf($[${ARG1} <= ${DB(${KEY})}]?log)
    same => n(return),Return() ; Return without logging
    same => n(log),Verbose(0,${ARG2})
    same => n,Return()
```

Discussion

The `chanlog GoSub()` routine takes two arguments:

ARG1
: A numeric log level

ARG2
: The log message

Channel logging using this routine will be sent to the Asterisk console at verbose level 0, meaning that they will show up when you want them to, regardless of the current *core set verbose* setting. This routine uses a different method, values in AstDB, to control what messages show up. See Table 1-1 for the AstDB entries read in by the `chanlog` routine. If the log level argument is less than or equal to one of these matched entries in the AstDB, then the message will be logged.

Table 1-1. chanlog AstDB entries

Family	Key	Description
ChanLog/	all	This setting is applied to all channels.
ChanLog/	channels/<device>	The `chanlog` routine will also look for an entry that matches the part of the channel name that comes before the -. For example, if the channel name is SIP/myphone-00112233, the routine will look for a key of channels/SIP/myphone.

Family	Key	Description
ChanLog/	channels/<full_channel>	The chanlog routine also checks for a match against the full channel name.

Once this routine has been added to your dialplan, you can start adding log statements to any extensions on your system. As an example, here is a typical company main menu that now has chanlog messages for each step:

```
exten => 6000,1,GoSub(chanlog,s,1(1,[${CHANNEL}] Entered main menu))
    same => n,Background(main-menu)
    same => n,WaitExten(5)

exten => 1,1,GoSub(chanlog,s,1(1,[${CHANNEL}] Pressed 1 for Sales))
    same => n,Goto(sales,s,1)

exten => 2,1,GoSub(chanlog,s,1(1,[${CHANNEL}] Pressed 2 for Technical Support))
    same => n,Goto(tech_support,s,1)

exten => 3,1,GoSub(chanlog,s,1(1,[${CHANNEL}] Pressed 3 for Customer Service))
    same => n,Goto(customer_service,s,1)

exten => 4,1,GoSub(chanlog,s,1(1,[${CHANNEL}] Pressed 4 for Training))
    same => n,Goto(training,s,1)

exten => 5,1,GoSub(chanlog,s,1(1,[${CHANNEL}] Pressed 5 for Directory))
    same => n,Directory()

exten => 0,1,GoSub(chanlog,s,1(1,[${CHANNEL}] Pressed 0 for Operator))
    same => n,Goto(operator,s,1)

exten => i,1,GoSub(chanlog,s,1(1,[$CHANNEL}] invalid "${INVALID_EXTEN}"))
    same => n,Goto(6000,1)

exten => t,1,GoSub(chanlog,s,1(1,[${CHANNEL}] Timed out waiting for digit))
    same => n,Goto(6000,1)
```

To see the output, we will first disable all other verbose messages from Asterisk and then turn on chanlog messages for all channels:

```
*CLI> core set verbose 0
*CLI> database put ChanLog all 1
```

Now when someone calls into the main menu, he will see messages like this at the Asterisk console:

```
[SIP/000411223344-000000b8] Entered main menu
[SIP/000411223344-000000b8] Pressed 4 for Training
```

See Also

For more information about other types of logging and monitoring of Asterisk systems, see Chapter 24, "System Monitoring and Logging," of *Asterisk: The Definitive Guide* (*http://oreilly.com/catalog/9780596517342/*) (O'Reilly).

CHAPTER 2

Call Control

2.0 Introduction

The recipes in this chapter focus on making and controlling phone calls in Asterisk. The rich set of possibilities for controlling calls is part of what makes Asterisk a telephony applications platform and not just another PBX.

2.1 Creating Call Limits Using Groups

Problem

You would like to implement custom call limits in the Asterisk dialplan.

Solution

Use the GROUP() and GROUP_COUNT() dialplan functions:

```
exten => _1NXXNXXXXXX,1,Set(GROUP(outbound)=myprovider)
    same => n,Set(COUNT=${GROUP_COUNT(myprovider@outbound)})
    same => n,NoOp(There are ${COUNT} calls for myprovider.)
    same => n,GotoIf($[${COUNT} > 2]?denied:continue)
    same => n(denied),NoOp(There are too many calls up already.  Hang up.)
    same => n,HangUp()
    same => n(continue),GoSub(callmyprovider,${EXTEN},1)
```

Discussion

In this example solution, we have shown how you could use the GROUP() and GROUP_COUNT() functions to limit the number of calls sent to a provider to no more than two calls at a time. You can think of using the GROUP() function as a way to apply a marker to a channel. GROUP_COUNT() is the method of getting a count of how many active channels are up with a given mark on them.

The argument provided to the GROUP() function is a category. In our example, the category used is outbound. The first line sets the channel's outbound group value to myprovider. On the next line, we set the COUNT variable to the number of channels that are marked as being in the myprovider group within the outbound category. The rest of the example demonstrates using this value in a conditional statement so that the dialplan can continue down two different paths depending on the number of channels that matched the GROUP_COUNT() criteria.

One reason that you might want to use this particular example is if you want to limit your exposure to fraudulent calls if an account on your system were to be compromised. An attacker would likely send as many calls through your system at a time as they could, so by limiting the number of calls that cost you money, you limit how many charges an attacker could rack up on you before you catch it.

GROUP() and GROUP_COUNT() can be used in a lot of other cases, too. By using a very similar approach to this example, you can also limit the number of calls that a specific user account is able to make at any given time. Another use would be to set a global call limit on the system.

See Also

The usage of GROUP() and GROUP_COUNT() comes up a number of times in *Asterisk: The Definitive Guide* (O'Reilly). References can be found in Chapters 13 ("Automatic Call Distribution (ACD) Queues"), 14 ("Device States"), 22 ("Clustering"), 23 ("Distributed Universal Number Discovery (DUNDi)"), and 26 ("Security").

2.2 Originating a Call Using the CLI

Problem

As an Asterisk system administrator, you would like to quickly originate a new call from the Asterisk command-line interface.

Solution

Use the *channel originate* CLI command. To connect a channel directly to an application:

```
*CLI> channel originate SIP/myphone application Playback demo-congrats
```

To connect a channel to an extension in the dialplan:

```
*CLI> channel originate SIP/myphone extension 1234@DialplanContext
```

Discussion

Originating calls is a fairly common task. The CLI version of this functionality is most useful when doing quick test calls while writing Asterisk dialplan. The examples above showed how to originate a call to a phone and connect it to something Asterisk. When testing, the use of a Local channel instead of a real phone is incredibly handy. For the purposes of quickly testing some dialplan logic, you can just create an extension that runs the `Wait()` application:

```
; /etc/asterisk/extensions.conf

[default]
exten => wait,1,Answer()
    same => n,Wait(300)
    same => n,Hangup()

exten => newexten,1,Verbose(1,I wonder if my CUT() works...)
    same => n,Set(VAR=one-two-three)
    same => n,Verbose(1,one = ${CUT(VAR,-,1)})
```

Now, to quickly test whether this bit of dialplan logic is working correctly, you can run the following:

```
*CLI> channel originate Local/wait@default extension newexten@default
I wonder if my CUT() works...
one = one
```

See Also

Recipe 2.3, Recipe 2.4, Recipe 2.5

2.3 Originating a Call Using the Dialplan

Problem

You would like to originate a call from the Asterisk dialplan.

Solution

Asterisk provides an application for originating calls from the dialplan. To originate a call and connect it to an application, you would do this:

```
exten => s,1,Originate(SIP/myphone,app,Playback,all-your-base)
```

Alternatively, you can originate a call and connect it to an extension in the dialplan:

```
exten => s,1,Originate(SIP/myphone,exten,default,s,1)
```

Discussion

The `Originate()` application takes up to 5 arguments. The first two are:

Tech/data
> This is the channel technology and associated data that the call will be originated to. The syntax is the same as is used with the commonly used `Dial()` application.

originate mode
> There are two originate modes: app and exten. The app originate mode is used to connect the originated call to an Asterisk application. If more complex call processing is desired, the exten originate mode can be used to connect the originated call to an extension in the dialplan.

The rest of the arguments depend on which originate mode is being used. In the case of the app originate mode, the arguments are:

application
> This is the dialplan application that will be answered when the dialed channel answers.

application arguments
> Any other arguments are passed to the application being executed.

If the originate mode is exten, the rest of the arguments are:

context
> This is the context in the dialplan that should be used to find the extension to run. This argument is required.

extension
> This is the extension in the dialplan that will be executed. If this argument is not specified, the s extension will be used.

priority
> This is the priority of the extension to execute. If this is not specified, dialplan execution will start at priority 1, which is almost always what you want.

There are a lot of cases where you might want to originate a call from the dialplan. One such example is related to the handling of paging. Perhaps you would like to write an extension that allows a caller to record and review a message to be played out through a paging system, but allow the caller to hang up after recording it, instead of doing the page live or requiring the caller to stay on the phone while the paging process finishes. The way you can accomplish this task is by first writing the dialplan that allows the caller to record something. Afterwards, the Originate() application will be used to trigger the paging process to begin. That way, even if the caller hangs up, the paging process will continue. Let's get on to the example:

```
[globals]
PHONES_TO_PAGE=SIP/phoneA&SIP/phoneB&SIP/phoneC


[paging]
exten => 500,1,Answer()
    same => n,Record(/tmp/page.wav)
    same => n,Originate(Local/pageplayback@paging,exten,paging,page,1)
    same => n,Hangup()

exten => page,1,Answer()
```

```
;
; This causes Polycom phones to auto-answer the call.
;
    same => n,SIPAddHeader(Alert-Info: Ring Answer)
    same => n,Page(${PHONES_TO_PAGE})
    same => n,Hangup()

exten => pageplayback,1,Answer()
    same => Playback(/tmp/page.wav)
```

In this example, a caller would dial 500 to initiate the paging process. The Record()
application would allow them to record their announcement to a file. The recording
will end and be saved to disk when the caller presses the # key on their phone. The
paging process can be canceled from this point by just hanging up. Next, the Origi
nate() application starts a new call. On one end of the call is a Local channel, which is
executing a short extension that just plays back the recording that was just left. The
other end of the call is an extension that uses the Page() application.

Before executing the Page() application, the SIPAddHeader() application is used to set
a header that will be added to outbound call requests to SIP phones since we want the
call to be automatically answered by the phones. This specific example will work for
Polycom phones. For examples of how to get other brands of phones to automatically
answer, see the "Parking and Paging" chapter of *Asterisk: The Definitive Guide*.

See Also

The "External Services" chapter of *Asterisk: The Definitive Guide* has a section on in-
tegrating Asterisk with calendar systems. One of the features provided by calendar
integration in Asterisk is the ability to have calls originated based on calendar events.
One of the really great examples in that book shows how to read information out of
the calendar event in the dialplan and then use the Originate() application to make
calls to all participants in the meeting.

Other recipes related to this one include Recipe 2.2, Recipe 2.4, and Recipe 2.5.

2.4 Originating a Call From a Call File

Problem

You would like to programmatically originate a call outside of Asterisk directly from
the Asterisk server in the simplest way possible.

Solution

Create a call file. Use your favorite text editor to create a file called *example.call* with
the following contents:

```
Channel: SIP/YourPhone
```

```
Context: outbound
Extension: 12565551212
Priority: 1
```

Once the file has been created, move it to the Asterisk spool directory:

```
$ mv example.call /var/spool/asterisk/outgoing/
```

Discussion

Call files are a great straightforward way to originate a call from outside of Asterisk. There are some things that are important to understand about call files if you choose to use them, though. First, you must understand the syntax and options that are allowed from within a call file. Beyond that, there are some nuances regarding how Asterisk processes call files that you should be aware of.

Every line in a call file is specified as a key/value pair:

```
key: value
```

The one line that is required in every call file is the specification of which channel to originate a call to:

```
Channel: Tech/data
```

There are two modes for all of the different methods of originating calls. The first mode is for when you connect a channel directly to an application. The second mode is for when you connect a channel to an extension in the dialplan. For the first mode, connecting to an application, you must specify which application and the arguments to pass to it:

```
Application: MeetMe
Data: 1234
```

For the other mode, connecting a channel to an extension in the dialplan, you must specify the context, extension, and priority in the call file:

```
Context: default
Extension: s
Priority: 1
```

The rest of the parameters that may be specified in call files are all optional, but some of them are incredibly useful:

Codecs: *ulaw, alaw, gsm*
> By default, Asterisk will allow the outbound channel created at the start of the call origination process choose whatever codec(s) it wants. If you would like to impose some limits on which codecs the channel may choose, you can specify them as a comma delimited list with the Codecs option.

MaxRetries: *2*
> If the outbound call to the specified channel fails, this is how many times the call will be retried. By default, the call will only be attempted one time.

RetryTime: *60*
> This option specifies the number of seconds to wait in between retries. The default is **300** seconds.

WaitTime: *30*
> This is the number of seconds to wait for an answer from the outbound call before considering it a failed attempt. By default, this is set to **45** seconds.

CallerID: *"Russell Bryant" <(256) 555-1212>*
> Specify the CallerID to use for the outbound call. By default, no CallerID information is provided for the outbound called made to Channel.

Account: *accountcode*
> Set the accountcode field on the outbound channel. By default, this field is not set at all. The accountcode is used in both Call Detail Record (CDR) and Channel Event Logging (CEL) processing. More information about CDR and CEL can be found in the "Monitoring and Logging" chapter of *Asterisk: The Definitive Guide*.

AlwaysDelete: *Yes*
> Always delete the call file when Asterisk is done processing it. Normally, once Asterisk has finished successfully making the call or has given up after the configured number of retries, the call file will be deleted. If this option is set to No and the timestamp of the file is modified before Asterisk finishes processing it to be some point in the future, the file will not be deleted. The result is that Asterisk will process this file again when the new time is reached.

Set: *CHANNELVAR=value*
> Set a channel variable on the outbound channel to a specified value.

Set: *FUNCTION(functionargs)=value*
> Set a dialplan function on the outbound channel to a specified value.

Archive: *Yes*
> Archive call files after processing them. If this option is set to Yes, instead of deleting the call file after processing, Asterisk will move it to the */var/spool/asterisk/outgoing_done/* directory. Before the file is moved, Asterisk will add a Status line to the file to indicate if the originated call was Completed, Expired, or Failed.

Now that we have covered all of the options that can be specified, the last bit of business is to make you aware of some of the important details about how Asterisk processes call files. Here are two critical issues you must keep in mind:

- Never create a call file in the */var/spool/asterisk/outgoing/* directory. Asterisk starts processing these files as soon as they are created. If you create it directly in this directory, Asterisk may read the contents before you are finished writing the file. Instead, always create the file somewhere else on the filesystem and move it into this directory when you are ready to make Asterisk aware of its presence.

```
# vi /tmp/example.call
# mv /tmp/example.call
/var/spool/asterisk/outgoing/
```

- The timestamp on a call file is important. If you set the timestamp to be in the future, Asterisk will not process the file until that time is reached.

See Also

Recipe 2.2, Recipe 2.3, and Recipe 2.5

2.5 Originating a Call From the Manager Interface

Problem

You would like to programmatically originate a call over a network connection to Asterisk.

Solution

Use the Originate action in the Asterisk Manager Interface (AMI):

```
Action: Originate
Channel: SIP/myphone
Exten: 6001
Context: LocalExtensions
Priority: 1
Timeout: 30000
CallerID: "Asterisk" <6000>
Async: true
```

Discussion

The Originate action in the AMI allows you to send a request over a TCP connection for Asterisk to make a call. This is the most popular method for originating calls from custom applications. The example provided in the solution starts by having Asterisk make a new call to *SIP/myphone*. If the phone does not answer within 30 seconds, the call will be aborted. If the call is answered, it is connected to extension *6001* in the *LocalExtensions* context in the dialplan.

Alternatively, the Originate action can be used to connect a channel directly to an application. By starting with the example provided in the solution, we can modify it to connect the called phone to conference bridge number 1234 without going through the dialplan:

```
Action: Originate
Channel: SIP/myphone
Application: MeetMe
Data: 1234
Timeout: 30000
CallerID: "Asterisk" <6000>
Async: true
```

There are a few more useful optional headers that can be provided with the Origi
nate action:

ActionID: <value>
> This is a custom value that will also be included in any responses to this request.
> It can be helpful in custom applications that may have many outstanding requests
> at any one time to ensure that responses are associated with the proper request.

Variable: NAME=VALUE
> This header can be specified multiple times in an Originate request. It will set
> channel variables on the outbound channel. This can also be used to set dialplan
> functions. Just use a function such as CDR(customfield) in the NAME portion of the
> header.

Account
> Specify the account code that will be placed in the CDR for this call.

For additional information about the Originate manager action included with your
version of Asterisk, use this command at the Asterisk CLI:

```
*CLI> manager show command Originate
```

See Also

For more information about the Asterisk Manager Interface, see the AMI chapter in
Asterisk: The Definitive Guide.

For recipes that are directly related to this one, see Recipe 2.2, Recipe 2.3, and Rec
ipe 2.4.

2.6 Using the FollowMe() Dialplan Application

Problem

You would like to call a series of phone numbers to attempt to locate a person when
their extension is dialed.

Solution

Use the FollowMe() application. First you must configure the application in */etc/asterisk/
followme.conf*:

```
;
; Configure what to do when FollowMe() is requested for Russell.
; We are going to try his desk phone first, and then try to call
; his cell phone.
;
[russell]
;
; FollowMe() will use this context in the Asterisk dialplan
```

```
; to make outbound calls.
;
context = trusted
;
; Call this number first.  Give up after 20 seconds.
;
number = 7101,20
;
; Call this number second.  Give up after 20 seconds.
;
number = 12565551212,20


[leif]
context = trusted
number = 7102,20
number = 12565559898,20
```

Now create extensions in the dialplan that will utilize the FollowMe() application to locate the called party:

```
[public_extensions]
exten => 7001,1,FollowMe(russell)

exten => 7002,1,FollowMe(leif)


[trusted]
exten => 7101,1,Dial(SIP/russell_desk)

exten => 7102,1,Dial(SIP/leif_desk)

exten => _1NXXNXXXXXX,1,Dial(SIP/${NUMBER}@outbound_provider)
```

Discussion

Admittedly, everything that the FollowMe() application does for you can be implemented directly in the dialplan. However, it takes more work. If this application does what you want, then save yourself the extra effort and just use FollowMe().

When the FollowMe() application executes, it is going to load the list of phone numbers that were configured in *followme.conf* and dial them in the order they were specified. When one of the outbound calls is answered, a prompt will be played requesting that the call be acknowledged by pressing a key. One of the main reasons for requiring acknowledgment is that it ensures that if the call is answered by voicemail that FollowMe() continues and tries other numbers.

See Also

There are some additional options available in the *followme.conf* file, but they are rarely used. For additional information on the current set of available options, see *configs/*

followme.conf.sample in the Asterisk source. For some additional information about the syntax of `FollowMe()` in the dialplan, check the documentation built in to Asterisk:

```
*CLI> core show application FollowMe
```

If you are interested in building similar functionality directly in the dialplan, see Recipe 2.7.

2.7 Building Find-Me-Follow-Me in the Dialplan

Problem

You would like to implement find-me-follow-me in the Asterisk dialplan.

Solution

```
[public_extensions]
;
; Find Russell.
;
exten => 7001,1,Progress()
    same => n,Playback(followme/pls-hold-while-try,noanswer)
    same => n,Dial(Local/7101@trusted,20,rU(ackcall^s^1))
    same => n,Dial(Local/12565551212@trusted,20,rU(ackcall^s^1))
    same => n,Playback(followme/sorry,noanswer)
    same => n,Hangup()


;
; Find Leif.
;
exten => 7002,1,Progress()
    same => n,Playback(followme/pls-hold-while-try,noanswer)
    same => n,Dial(Local/7102@trusted,20,rU(ackcall^s^1))
    same => n,Dial(Local/12565559898@trusted,20,rU(ackcall^s^1))
    same => n,Playback(followme/sorry,noanswer)
    same => n,Hangup()


[trusted]
exten => 7101,1,Dial(SIP/russell_desk)

exten => 7102,1,Dial(SIP/leif_desk)

exten => _1NXXNXXXXXX,1,Dial(SIP/${NUMBER}@outbound_provider)


[ackcall]
exten => s,1,Background(followme/no-recording&followme/options)
    same => n,WaitExten(5)
    same => n,Set(GOSUB_RESULT=BUSY)

exten => 1,1,NoOp()
```

```
exten => 2,1,Set(GOSUB_RESULT=BUSY)

exten => i,1,Set(GOSUB_RESULT=BUSY)
```

Discussion

This example implements the same functionality provided by the FollowMe() application shown in Recipe 2.6. The most important difference is that if you want to tweak any of the behavior, it is much easier to do so in the Asterisk dialplan as opposed to modifying the C code of the FollowMe() application.

Here is what happens when someone dials 7001 to reach Russell:

```
exten => 7001,1,Progress()
```

The Progress() application is used for telephony signaling. We do not want to answer the call yet, but we want to start playing audio. This is often referred to as early media in a phone call. The use of Progress() in the dialplan maps to sending a 183 Session Progress request in SIP:

```
same => n,Playback(followme/pls-hold-while-try,noanswer)
```

Now that we have told the calling channel to expect early media, we are going to play a prompt without answering the call. This prompt says "Please hold while I try to locate the person you are calling":

```
same => n,Dial(Local/7102@trusted,20,rU(ackcall^s^1))
```

There is a lot packed into this line. First, we are making an outbound call to a Local channel. This local extension just makes an outbound call to a single SIP device. We could have simply dialed that SIP device directly and the behavior would have been the same. However, the use of the Local channel is more in line with how the FollowMe() application works.

In the *followme.conf* file, you configure phone numbers, not devices, for the application to call. We specify a 20 second timeout for this outbound call attempt. Finally, we set a couple of options. The first option is r, which ensures that Asterisk generates a ringback tone. We want this because we have already indicated to the caller that we will be providing early media.

An alternative to this would be to use the m option of Dial(), which would provide hold music instead of ringback. Finally, we use the U option, which executes a GoSub() routine on the called channel after it answers, but before connecting it to the inbound channel. The routine we are using is ackcall, which gives the called channel the option of whether to accept the call. We will come back to the implementation of the ack call routine shortly.

Now let's look at the next step:

```
same => n,Dial(Local/12565551212@trusted,20,rU(ackcall^s^1))
```

This step is identical to the last one except that it is calling a different number. You could have as many of these steps as you would like.

The next line is:

```
same => n,Playback(followme/sorry,noanswer)
```

If the caller makes it this far in the dialplan, Asterisk was not able to find the called party at any of the configured numbers. The prompt says "I'm sorry, but I was unable to locate the person you were calling".

Finally, we have:

```
same => n,Hangup()
```

As you might have guessed, this hangs up the call. Now let's go back to the implementation of the ackcall GoSub() routine that is executed by the Dial() application:

```
[ackcall]
exten => s,1,Background(followme/no-recording&followme/options)
    same => n,WaitExten(5)
    same => n,Set(GOSUB_RESULT=BUSY)

exten => 1,1,NoOp()

exten => 2,1,Set(GOSUB_RESULT=BUSY)

exten => i,1,Set(GOSUB_RESULT=BUSY)
```

The execution of this routine begins at the s extension. The Background() application is going to play two prompts while also waiting for a digit to be pressed. The called party will hear "You have an incoming call. Press 1 to accept this call, or 2 to reject it." After the prompts finish playing, the WaitExten() application will wait an additional five seconds for a key to be pressed. At this point, there are four different cases that may occur:

The called party does nothing.
> In this case, the s extension will continue to the next step and the GOSUB_RESULT variable will be set to BUSY. This makes the Dial() application act like this outbound call attempt returned a busy response.

The called party presses 1 to accept.
> The call will jump to the 1 extension which does nothing. Control will return to the Dial() application and the caller and callee will be bridged together.

The called party presses 2 to reject.
> The call will jump to the 2 extension which sets the GOSUB_RESULT variable to BUSY. Dial() will treat this outbound call attempt as busy.

The caller presses a different key.
> Asterisk will look for an extension that matches the key that was pressed but will not find one. The call will instead go to the i extension, which stands for invalid.

The GOSUB_RESULT variable will be set to BUSY and Dial() will treat this outbound call attempt as busy.

See Also

Consider using the FollowMe() application as discussed in Recipe 2.6 if you do not require the level of customization required by implementing this functionality in the dialplan.

For more information about Local channels, see Chapter 10, "Deeper into the Dialplan," in *Asterisk: The Definitive Guide*.

2.8 Creating a Callback Service in the Dialplan

Problem

After a call has failed due to the destination being busy or otherwise unavailable, you would like to give the caller the option of being automatically called back when the destination becomes available.

Solution

Add the following options to the phone configuration sections of */etc/asterisk/sip.conf*:

```
[phone1]
...
cc_agent_policy = generic
cc_monitor_policy = generic

[phone2]
...
cc_agent_policy = generic
cc_monitor_policy = generic
```

Next, add the *30 and *31 extensions to your dialplan:

```
[phones]
;
; The extensions for dialing phones do not need to be changed.
; These are just simple examples of dialing a phone with a
; 20 second timeout.
;
exten => 7101,1,Dial(SIP/phone1,20)
    same => n,Hangup()

exten => 7102,1,Dial(SIP/phone2,20)
    same => n,Hangup()


;
; Dial *30 to request call completion services for the last
; call attempt.
;
```

```
exten => *30,1,CallCompletionRequest()
    same => n,Hangup()


;
; Dial *31 to cancel a call completion request.
;
exten => *31,1,CallCompletionCancel()
    same => n,Hangup()
```

Discussion

Call Completion Supplementary Services (CCSS) is a new feature in Asterisk 1.8. It allows you to request that Asterisk call you back after an unanswered or busy call attempt. In this example, we have used the generic agent and monitor policies. This method of configuration is the easiest to get working but only works for calls between phones connected to the same Asterisk system. The agent is the part of the system that operates on behalf of the caller requesting call completion services. The monitor is the part of the system that is in charge of monitoring the device (or devices) that were called to determine when they become available.

Using this dialplan, let's go through an example where CCSS is used. Start by having 7001 call 7002, but let the call time out after the configured 20 seconds. In this case, the call has failed due to no response. The caller, 7001, can now request CCSS by dialing *30. This is referred to as Call Completion No Response (CCNR). You can verify the CCNR request at the Asterisk CLI:

```
*CLI> cc report status
1 Call completion transactions
Core ID        Caller                        Status
--------------------------------------------------------------------------
20             SIP/phone1                    CC accepted by callee
               |-->7102@phones
               |-->SIP/phone2(CCNR)
```

At this point, Asterisk is using its generic monitor implementation to wait for SIP/phone2 to become available. In the case of CCNR, it determines availability by waiting for the phone to make a call. When that call ends, Asterisk will initiate a call between SIP/phone1 and SIP/phone2 and the CCNR request will have been completed.

Another scenario is Call Completion Busy Subscriber (CCBS). This is the case when the called party is already on the phone. The process of requesting call completion services in this scenario is the same as before. The generic monitor will determine availability by waiting for the call that device is on to end.

Asterisk also supports extending CCSS across multiple servers using either SIP or ISDN (specifically ISDN in Europe). However, the configuration and operation of the protocol specific methods is outside the scope of this recipe.

See Also

The Asterisk project wiki, *http://wiki.asterisk.org/*, discusses CCSS.

2.9 Hot-Desking with the Asterisk Database

Problem

You need people to be able to log in to any device and accept calls at that location, a feature known as hot-desking.

Solution

```
[HotDesking]
; Control extension range using pattern matches

; Login with 71XX will logout existing extension at this location
; and log this device in with new extension.
; Logoff with 7000 from any device.
;

exten => 7000,1,Verbose(2,Attempting logoff from device ${CHANNEL(peername)})
   same => n,Set(PeerName=${CHANNEL(peername)})
   same => n,Set(CurrentExtension=${DB(HotDesk/${PeerName})})
   same => n,GoSubIf($[${EXISTS(${CurrentExtension})}]?
subDeviceLogoff,1(${PeerName},${CurrentExtension}):loggedoff)
   same => n,GotoIf($[${GOSUB_RETVAL} = 0]?loggedoff)
   same => n,Playback(an-error-has-occurred)
   same => n,Hangup()
   same => n(loggedoff),Playback(silence/1&agent-loggedoff)
   same => n,Hangup()


exten => _71XX,1,Verbose(2,Attempting to login device ${CHANNEL(peername)}
to extension ${EXTEN:1})
   same => n,Set(NewPeerName=${CHANNEL(peername)})
   same => n,Set(NewExtension=${EXTEN:1})

; Check if existing extension is logged in for this device (NewPeerName)
; -- If existing extension exists (ExistingExtension)
;     -- get existing device name
;        -- If no existing device
;           -- (login) as we'll overwrite existing extension for this device
;        -- If existing device name
;           -- logoff ExistingExtension + ExistingDevice
;              -- Goto check_device -------------------------------------+
; -- If no existing extension exists                                     |
;     -- Check if existing device is logged in for this extension        |
;        (NewExtension) <-----------------------------------------------+
;           -- If existing device exists
;              -- Get existing extension
;                 -- If extension exists
;                    -- Logoff Device + Extension
;                       -- Login
;                 -- If no extension exists
;                    -- Remove device from AstDB
;                       -- Login
```

```
;          -- If no device exists for NewExtension
;              -- Login

; Tests:
; * Login 100 to 0000FFFF0001
; * Login 101 to 0000FFFF0001 (Result: Only 101 logged in)
; * Login 101 to 0000FFFF0002 (Result: Only 101 logged in to new location)
; * Login 100 to 0000FFFF0001 (Result: Both 100 and 101 logged in)
; * Login 100 to 0000FFFF0002 (Result: Only 100 logged into 0000FFFF0002
;                                      -- change locations)
; * Login 100 to 0000FFFF0001 (Result: Only 100 logged in)

    same => n,Set(ExistingExtension=${DB(HotDesk/${NewPeerName})})
    same => n,GotoIf($[${EXISTS(${ExistingExtension})}]?get_existing_device)

    same => n(check_device),NoOp()
    same => n,Set(ExistingDevice=${DB(HotDesk/${NewExtension})})
    same => n,GotoIf($[${EXISTS(${ExistingDevice})}]?get_existing_extension)
    same => n,NoOp(Nothing to logout)
    same => n,Goto(login)

    same => n(get_existing_device),NoOp()
    same => n,Set(ExistingDevice=${DB(HotDesk/${ExistingExtension})})
    same => n,GotoIf($[${ISNULL(${ExistingDevice})}]?login)
    same => n,GoSub(subDeviceLogoff,1(${ExistingDevice},${ExistingExtension}))
    same => n,GotoIf($[${GOSUB_RETVAL} = 0]?check_device)
    same => n,Playback(silence/1&an-error-has-occurred)
    same => n,Hangup()

    same => n(get_existing_extension),NoOp()
    same => n,Set(ExistingExtension=${DB(HotDesk/${ExistingDevice})})
    same => n,GoSubIf($[${EXISTS(${ExistingExtension})}]?
subDeviceLogoff,1(${ExistingDevice},${ExistingExtension}):remove_device)
    same => n,GotoIf($[${GOSUB_RETVAL} = 0]?loggedoff)
    same => n,Playback(silence/1&an-error-has-occurred)
    same => n,Hangup()
    same => n(remove_device),NoOp()
    same => n,Set(Result=${DB_DELETE(HotDesk/${ExistingDevice})})
    same => n,Goto(loggedoff)

    same => n(loggedoff),Verbose(2,Existing device and extensions have
been logged off prior to login)
    same => n(login),Verbose(2,Now logging in extension ${NewExtension}
to device ${NewPeerName})
    same => n,GoSub(subDeviceLogin,1(${NewPeerName},${NewExtension}))
    same => n,GotoIf($[${GOSUB_RETVAL} = 0]?login_ok)
    same => n,Playback(silence/1&an-error-has-occurred)
    same => n,Hangup()

    same => n(login_ok),Playback(silence/1&agent-loginok)
    same => n,Hangup()

exten => subDeviceLogoff,1,NoOp()
    same => n,Set(LOCAL(PeerName)=${ARG1})
    same => n,Set(LOCAL(Extension)=${ARG2})
```

```
    same => n,ExecIf($[${ISNULL(${LOCAL(PeerName)})} |
${ISNULL(${LOCAL(Extension)})}]?Return(-1))
    same => n,Set(PeerNameResult=${DB_DELETE(HotDesk/${LOCAL(PeerName)})})
    same => n,Set(ExtensionResult=${DB_DELETE(HotDesk/${LOCAL(Extension)})})
    same => n,Return(0)

exten => subDeviceLogin,1,NoOp()
    same => n,Set(LOCAL(PeerName)=${ARG1})
    same => n,Set(LOCAL(Extension)=${ARG2})
    same => n,ExecIf($[${ISNULL(${LOCAL(PeerName)})} |
${ISNULL(${LOCAL(Extension)})}]?Return(-1))
    same => n,Set(DB(HotDesk/${LOCAL(PeerName)})=${LOCAL(Extension)})
    same => n,Set(DB(HotDesk/${LOCAL(Extension)})=${LOCAL(PeerName)})
    same => n,Set(ReturnResult=${IF($[${DB_EXISTS(HotDesk/${LOCAL(PeerName)})}
& ${DB_EXISTS(HotDesk/${LOCAL(Extension)})}]?0:-1)})
    same => n,Return(${ReturnResult})
```

Discussion

Hot-desking is a fairly common feature that is gathering increased traction as Asterisk systems are deployed because of the inherent flexibility the dialplan provides. Older, traditional PBX systems apply an extension number to either a line on the system or with a device itself. With Asterisk, we have the ability to apply dialplan logic and information stored in a local database (or external database) to determine where an extension rings. We could easily develop a system where an extension number does nothing but ring a cell phone, or a combination of devices (like in a paging system, or a group of sales agents).

In the dialplan provided for this example of hot-desking, we've allowed people to log in to any device by dialing **71XX** where **1XX** is the person's extension number in the range 100 through 199. To log out the extension from the device, the user simply dials **7000** from the device to log out from. While it has made the dialplan and logic more complicated, the dialplan also takes into account other extensions already logged into a device someone wants to log in to, and automatically logs them out first. Additionally, if we were logged into another device previously and didn't log out before changing locations, the dialplan will log the extension out from the other device first before logging it into the new location.

No external scripts or databases[*] have been employed, which helps demonstrate the flexibility of Asterisk. Let's look at what happens when we log in an extension to a device that has no existing extension logged in.

First we do our checks as described by the logic flow in the comment block in the code. Figure 2-1 shows the visual representation of what we're checking for before logging the extension in.

---

[*] Although using an external database may actually have simplified the logic.

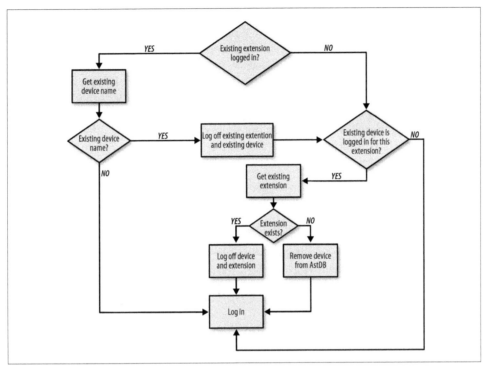

Figure 2-1. Login check logic

You'll need to be aware we haven't added any logic to authenticate callers. This may not be necessary, but, if so, you can add some additional
logic using one of the caller authentication recipes found in Chapter 1.
Additionally, we haven't added any prompts notifying the callers that
an existing extension is logged in prior to logging them out, as we wanted to keep the fundamental logic of the hot-desking application so you
have a base to work with.

To log in extension 100, we'll dial **7100**, which will place the appropriate information
into the AstDB: two rows located within the HotDesk family. After you've logged in,
you can see the entries in the database by entering **database show HotDesk** from the
Asterisk console:

```
*CLI> database show HotDesk
/HotDesk/0000FFFF0001                        : 100
/HotDesk/100                                 : 0000FFFF0001
2 results found.
```

The two entries are to provide a link between an extension→device and device→ extension. When logging off a device we'll need to know what extension to remove from the database, but when dialing an extension we'll need to know what device to call.[†] The flowchart in Figure 2-1 shows how we've performed our checks to help account for various situations. If we log in to a device, we have to make sure there wasn't another extension already logged in, and if so, log it out first. Another situation is where we were logged into another device and moved somewhere else, and need to move locations, which means we have to log out our extension from the other device first. And because we have multiple entries associated with each login, we have to verify that we've removed and modified both entries.

The following tests were performed to validate the logic used:

- Login 100 to 0000FFFF0001
- Login 101 to 0000FFFF0001 (Result: Only 101 logged in)
- Login 101 to 0000FFFF0002 (Result: Only 101 logged in to new location)
- Login 100 to 0000FFFF0001 (Result: Both 100 and 101 logged in)
- Login 100 to 0000FFFF0002 (Result: Only 100 logged into 0000FFFF0002—change locations)
- Login 100 to 0000FFFF0001 (Result: Only 100 logged in)

Using these tests, you can then step through the logic and look at the dialplan to understand all the checks going on, and validate that things are working as they should.

 It is possible that if two people attempt to log in to the same extension at the same time, or if someone else is logging into a device that was previously logged into by an extension moving locations (and attempting to log in at the same time as the other person) that the database could get out of sync. No locking has been performed here in order to keep the logic as clean and simple as possible. If there exists a strong possibility of people changing locations and logging in and out often on top of each other, then you may wish look into adding dialplan locking, which can be done using the LOCK() and UNLOCK() dialplan functions.

In more than one situation, the promise of hot-desking being just one more feature of the system is what eventually convinced a company to go with Asterisk because hot-desking was the killer application they were looking for.

---

† Because the AstDB is based on the usage of a family/key relationship, we need two entries and to keep them synchronized. This is a good reason why using an external relational database could actually simplify the hot-desking logic.

See Also

Recipe 1.4, Recipe 1.1, Recipe 1.2. Many of the concepts used in this dialplan stem directly from Chapter 10, "Deeper Into The Dialplan," in *Asterisk: The Definitive Guide* (Expressions and Variable Manipulation, Dialplan Functions, Conditional Branching, GoSub, and Using the Asterisk Database).

CHAPTER 3

Audio Manipulation

3.0 Introduction

The recipes in this chapter are designed to help you with the injection of audio into and the monitoring of channels in your Asterisk environment. Many of the recipes focus on a particular aspect but can be built up or modified using the skills learned in other recipes in this book.

3.1 Monitoring and Barging into Live Calls

Problem

As the manager of a call center, you need to be able to listen in on calls to help with training new employees.

Solution

The most simple solution is to simply connect to any active channel using the ChanSpy() application, which then provides you the ability to flip through active channels using DTMF. The b option means to only listen to actively bridged calls:

```
[CallCenterTraining]
exten => 500,1,Verbose(2,Listening to live agents)
    same => n,ChanSpy(,b)
```

If you only want to spy on certain channels, you can use the chanprefix option to control which types of channels you want to listen to. So, if we just want to listen to SIP channels involved in bridged calls, we would do this:

```
[CallCenterTraining]
exten => 500,1,Verbose(2,Listening to live agents)
    same => n,ChanSpy(SIP,b)
```

Discussion

The default DTMF keys for controlling ChanSpy() are as follows:

\#

    Cycles through the volume level

*

    Stop listening to the current channel and find another one to listen to

There are a lot more options for ChanSpy() and ways to use it in your dialplan. With some creativity, ChanSpy() can be used in many situations it wasn't necessarily designed for (see Recipe 3.4). Not only can ChanSpy() be used to listen to the conversation between channels, but you can also speak to a single channel where only one person can hear you, referred to as *whispering*.

Whispering to a channel is commonplace in situations where a manager of a call center is training an employee, and needs to listen to an agent during the call. To enable whispering to channels that are being spied on, use the w option:

```
[CallCenterTraining]
exten => 500,1,Verbose(2,Listening to live agents with whisper)
    same => n,ChanSpy(,bw)
```

Of course, we're going to be looking for some finer grain control for who we're listening to. Perhaps we have several groups of people (multiple campaigns, different products, etc.) we want to separate and listen to. One of these groups could simply be the training group. New agents are all placed into the training group which makes it easy to scan and listen to calls while providing help where necessary. To do this, we need to associate channels with a spygroup using the SPYGROUP channel variable. By setting the channel variable for a particular channel, ChanSpy() can then be directed to listen only to channels in a particular spygroup.

As we show in Recipe 3.7, we need to make sure the SPYGROUP channel variable is set on the channel we want to spy on and whisper to. In order to do this, we need to use the U() option of Dial(), which will execute a subroutine which sets the channel variable on the called channel verses the calling channel. While setting the SPYGROUP channel variable on the calling channel would still get us the ability to listen to calls like we want, the whispering would be performed to the wrong channel (i.e., the customer would be listening to the manager speak, not the agent):

```
[InboundCallsToAgents]
exten => 100,1,Goto(start,1)

exten => start,1,Verbose(2,Placing a call to an agent)
same => n,Dial(SIP/0000FFFF0001,15,U(subSetSpyGroup^training))
same => n,Hangup()

[subSetSpyGroup]
exten => s,1,Verbose(2,Setting spygroup)
```

```
same => n,Set(SPYGROUP=${ARG1})
same => n,Return()
```

We can make sure the channel variable SPYGROUP was set on the correct channel by using the **core show channel** CLI command at the Asterisk console. In our example, the device 0000FFFF0004 placed a call to 0000FFFF0001 using the dialplan we just wrote. By checking the channel variables set on the call, we can verify our dialplan is working correctly:

```
*CLI> core show channels
Channel                Location             State   Application(Data)
SIP/0000FFFF0004-000   start@InboundCalls:2 Up      Dial(SIP/0000FFFF0001,15,U(sub
SIP/0000FFFF0001-000   s@app_dial_gosub_vir Up      AppDial((Outgoing Line))
2 active channels
1 active call
2 calls processed
```

Having verified the direction of the call (0000FFFF0004 used the Dial() application which is connected to 0000FFFF0001), lets look at the channel variables set for the call:

```
*CLI> core show channel SIP/0000FFFF0004-TAB
...[snip]...
BRIDGEPVTCALLID=209ee58508fb319a0a2f615030316c28@172.16.0.161:5060
BRIDGEPEER=SIP/0000FFFF0001-00000001
DIALEDPEERNUMBER=0000FFFF0001
DIALEDPEERNAME=SIP/0000FFFF0001-00000001
DIALSTATUS=ANSWER
...[snip]...
```

We've snipped out a lot of text, but what we're looking for is to make sure the calling channel doesn't have the SPYGROUP set for it as we're expecting it to be enabled on the channel being called (the agent). Using the same technique we can verify that:

```
*CLI> core show channel SIP/0000FFFF0001-TAB
...[snip]...
BRIDGEPVTCALLID=ODIxNTAyNGUyNWViZmM5NGIyOGY1ZTVjYTQ1N2ExNTI.
BRIDGEPEER=SIP/0000FFFF0004-00000000
GOSUB_RETVAL=
SPYGROUP=training
...[snip]...
```

Having verified our channel variable has been set on the correct channel, lets create the dialplan that will allow our manager to listen and whisper to the channels in the training group:

```
[CallCenterTraining]
exten => 500,1,Verbose(2,Listening to live agents)
    same => n,ChanSpy(SIP,bwg(training))
```

With minor modifications to our existing ChanSpy() code, we can now listen to only channels in the training group (option g(training)) and whisper to them (option w). This same technique could be applied to only listening to specified extensions or callers by placing only a single channel into the spygroup.

Some additional options which you may find useful include (to see all available options, use **core show application ChanSpy** from the Asterisk CLI):

d

Override the standard DTMF actions and instead use DTMF to switch between the following modes:

4

Spy mode

5

Whisper mode

6

Barge mode

E

Exit when the spied on channel is disconnected.

q

Quiet. Don't play a beep or speak the channel being spied on prior to listening.

S

Stop when there are no more channels to spy on.

See Also

Recipe 3.4, Recipe 1.4, Recipe 1.5, Recipe 1.6, Recipe 3.7

3.2 Growing Your Company With PITCH_SHIFT()

Problem

You want to appear as a larger company to the outside world by manipulating the pitch of audio when dialing certain extensions.

Solution

```
exten => 100,1,Verbose(2,${CALLERID(all)} is calling reception)
    same => n,Set(CALLERID(name)=RCP:${CALLERID(name)})
    same => n,Set(PITCH_SHIFT(tx)=high)
    same => n,Dial(SIP/0000FFFF0001,30)
    same => n,Voicemail(reception@company,u)
    same => n,Hangup()

exten => 200,1,Verbose(2,${CALLERID(all)} is calling sales)
    same => n,Set(CALLERID(name)=SLS:${CALLERID(name)})
    same => n,Set(PITCH_SHIFT(tx)=low)
    same => n,Dial(SIP/0000FFFF0001,30)
    same => n,Voicemail(sales@company,u)
    same => n,Hangup()
```

Discussion

Using `PITCH_SHIFT()` you can modify the transmitted and/or received audio pitch either up or down. By changing the pitch of audio, you can sound like different people of the company. In our example we've used the transmission modifier (`tx`) of `PITCH_SHIFT()` to raise the pitch of the transmitted voice when extension 100 is dialed. Similarly we've modified the transmitted pitch to be lower when dialing extension 200. In order to know which department is being called, we've modified the callerID name by prepending either `RCP` for reception or `SLS` for sales.

For both extensions we've dialed the same device within the company. However, depending on which extension was dialed, we've configured separate voicemail boxes. When setting up the voicemail greetings, you'll need to modify the transmitted audio using `PITCH_SHIFT()` prior to calling `VoicemailMain()` so that your voice sounds the same across answered calls and the voicemail greeting.

`PITCH_SHIFT()` contains several shorthands for modifying pitch:

highest
　　Pitch is raised one full octave.

higher
　　Pitch is raised higher.

high
　　Pitch is raised.

low
　　Pitch is lowered.

lower
　　Pitch is lowered more.

lowest
　　Pitch is lowered one full octave.

In addition to the shorthands, you can pass the floating point number between `0.1` and `4.0`. A value of `1.0` has no effect on the pitch. A number lower than `1.0` lowers the pitch and number greater than `1.0` will raise the pitch. This can provide some fine grained tuning of the pitch so that you don't sound like a robot or Mickey Mouse (which simply has the effect of being funny, and also makes you difficult to understand).

Testing can be done either by dialing another extension at your desk, or making use of the `Record()` application and then listening to the audio recorded. With some fine tuning you can grow your company without hiring additional employees!

3.3 Injecting Audio into a Conference Bridge

Problem

You need to play an audio file into a conference room.

Solution

```
[ConferenceAudio]
; Users would join the conference at extension 100
exten => 100,1,Goto(start,1)

; Trigger audio playback with extension 200
exten => 200,1,Originate(Local/inject@ConferenceAudio/n,exten,
ConferenceAudio,quiet_join,1)

; Users join the conference here
exten => start,1,NoOp()
    same => n,Answer()
    same => n,MeetMe(31337,d)
    same => n,Hangup()

; Use a couple flags to quietly enter the conference
exten => quiet_join,1,NoOp()
    same => n,Answer()
    same => n,MeetMe(31337,dtq)

; This triggers the file to be played into the conference
exten => inject,1,NoOp()
    same => n,Playback(silence/1&tt-weasels)
    same => n,Hangup()
```

Discussion

Playing audio into a conference is really quite a straightforward process. While there are several ways of triggering the audio playback through a call origination process (as shown in the See Also section) we've utilized the dialplan origination method for our example. Using the `Originate()` application, we've used a `Local` channel that plays back an audio file and connected it to the conference bridge using `MeetMe()` by connecting via a dialplan extension[*]. When we dial extension 200, the `Originate()` application creates a `Local` channel that executes the `inject` extension in the `ConferenceAudio` context. Once the `Local` channel is created, it is connected to the `quiet_join` extension located in the `ConferenceAudio` context which then connects to the `MeetMe()` application.

Once all these pieces are connected together, the audio is played into the conference bridge. We've set 3 flags on `MeetMe()` within the `quiet_join` extension. They are as follows:

---

[*] We could have just as easily used `ConfBridge()` as well. It just depends on your requirements.

---

d

Dynamically create the conference (don't use *meetme.conf*).

t

Talk-only—don't listen to audio. Not necessary, but could save on resources.

q

Quiet join—don't play join sounds when connecting to the conference (suppresses the beep audio file).

Our example is a barebones implementation, but with further development, this could be expanded to limit conference lengths with audio being injected periodically to let the participants know how long is left. Or for weekly meetings, agenda items could be injected into the conference to make sure things progress during the allocated meeting time much like is done on shows like Pardon The Interruption.

See Also

Recipe 3.4, Recipe 2.4, Recipe 2.3, Recipe 2.5, Recipe 3.7

3.4 Triggering Audio Playback into a Call Using DTMF

Problem

During a call you want to play back audio to a caller when they send a particular sequence of DTMF tones.

Solution

First we start with the creation of a new *applicationmap*:

```
; features.conf
[applicationmap]
play_message => #1,self/caller,Macro(PlayMessage)
```

Then we create our `Macro()` for triggering the message injection in *extensions.conf*:

```
[macro-PlayMessage]
exten => s,1,NoOp()
    same => n,Set(EncodedChannelToPass=${URIENCODE(${DYNAMIC_PEERNAME})})
    same => n,Originate(Local/spy-${EncodedChannelToPass}@whisper-channel/n,
exten,whisper-channel,audio,1)
```

Our `Macro()` calls via a `Local` channel another context which does the audio injection:

```
[whisper-channel]
exten => _spy-.,1,NoOp()
    same => n,Set(EncodedChannel=${CUT(EXTEN,-,2-3)})
    same => n,Set(GROUP(whisper-channel)=${EncodedChannel})
    same => n,ExecIf($[${GROUP_COUNT(${EncodedChannel}@
whisper-channel)} > 1]?Hangup())
    same => n,Set(ChannelToSpy=${URIDECODE(${EncodedChannel})})
```

```
    same => n,ChanSpy(${ChannelToSpy},wsqEB)
    same => n,Hangup()

exten => audio,1,NoOp()
    same => n,Answer()
    same => n,Wait(0.4)
    same => n,Set(VOLUME(TX)=-4)

; One example is to return current cost of the call. A lookup to a database or
; webservice would be required to make the data dynamic.
;
    same => n,SayNumber(7)
; letters/dollar vs. digits/dollars (plural)
    same => n,Playback(digits/dollars)
    same => n,SayNumber(48)
    same => n,Playback(cents)
    same => n,Hangup()
```

And then to enable it we need to add the following line into the dialplan that is executed by the channel that will be triggering the audio playback:

```
exten => _2XX,1,NoOp()
    same => n,Set(DYNAMIC_FEATURES=play_message)
    same => n,Dial(SIP/${EXTEN},30)
...
```

Discussion

In this solution we've developed a system using the *features.conf* file to trigger playback of audio via DTMF[†] to a channel without disconnecting the existing call. We've configured the feature called play_message (which could be any name not already in use), and assigned the #1 DTMF key combination to be the trigger. The option defined as self/caller means that the channel placing the call will be the one able to trigger the playback, and will also be the channel that hears the audio that is injected into the call. This is all done via the PlayMessage macro we'll be defining in the dialplan.

In the macro-PlayMessage context, we've assigned the name of the channel we'll be injecting audio to the EncodedChannelToPass channel variable. The channel name is obtained from the DYNAMIC_PEERNAME channel variable which was set when the play_message feature was trigged via the #1 DTMF sequence. We've used the URIENCODE() dialplan function to make it easier to pass the value of the channel using the Local channel inside the Originate() application. URIENCODE() is required as there will be front-slash (/) in the channel name, which would break the format expected by the Originate() command.

The Originate() application is used to trigger dialplan for playing back the audio prompt. The values for where to play the audio are passed via the spy-<*channel name*> extension in the whisper-channel context. The value passed to us through the

---

† Dual-Tone Multi-Frequency, aka Touch-Tone.

---

extension was URIENCODE()'d, so we need to decode it. The value passed to URIEN CODE() comes from the CUT() function, which takes the extension and separates on the hyphen, only passing the second and third fields, and assigning them to the ChannelTo Spy variable. We are essentially cutting off the "spy-" part of the extension, as it contains no useful information. The rest of the extension is the channel name we need to spy on, as passed to us from the Originate() command in the PlayMessage macro.

Using the ChannelToSpy channel variable, we then inject audio into the channel using the ChanSpy() application to whisper to the channel. Here we look back at the Origi nate() line which had a Local channel passed to it as one of the values, with the re- maining values being exten, whisper-channel, audio, and 1. The remaining values are telling the Originate() command to connect the Local channel with the extension audio within the whisper-channel context, starting at the first priority. The audio extension is what will contain the instructions for what audio to be played into the channel.

The flags passed to the ChanSpy() application are as follows:

w

    Whisper to the channel (allow us to listen and speak)

s

    Silent; don't play the beep

q

    Quiet; don't play the channel type

E

    Exit after the bridge is closed

B

    Barge in

Once the Originate() application has connected the ChanSpy() application to the re- quested channel, then we need to play back some audio. The audio to be played into the channel is handled via the audio extension within the whisper-channel context. Because we want to play the audio over the top to give the caller some information, but don't want to interrupt the call, we've lowered the volume of the audio using the VOLUME() function prior to playing any audio.

We've defined some static data as an example of what audio playback might be like. Of course, you'd need to add a call above that to make the data dynamic, which could be done via *func_odbc* (if the data was in a relational database), via CURL() (if returned from a webservice), or any other number of ways in which the data could be looked up and returned.

See Also

Recipe 3.3, Recipe 3.7, Recipe 3.1

3.5 Recording Calls from the Dialplan

Problem

You would like to enable call recording from the Asterisk dialplan.

Solution

Use the MixMonitor() application:

```
exten => 7001,1,MixMonitor(${UNIQUEID}.ulaw)
    same => n,Dial(SIP/myphone)
```

Discussion

MixMonitor() records the audio from both directions of the phone call and writes it to a file on disk in one of the audio formats that Asterisk supports. You can see a list of the file formats that your version of Asterisk supports at the Asterisk CLI. The Extensions column identifies which file format extensions can be used in the recording filename:

```
*CLI> core show file formats

Format      Name        Extensions
------      ----        ----------
gsm         wav49       WAV|wav49
slin16      wav16       wav16
slin        wav         wav
adpcm       vox         vox
slin16      sln16       sln16
slin        sln         sln|raw
siren7      siren7      siren7
siren14     siren14     siren14
g722        g722        g722
ulaw        au          au
alaw        alaw        alaw|al|alw
ulaw        pcm         pcm|ulaw|ul|mu|ulw
ilbc        iLBC        ilbc
h264        h264        h264
h263        h263        h263
gsm         gsm         gsm
g729        g729        g729
g726        g726-16     g726-16
g726        g726-24     g726-24
g726        g726-32     g726-32
g726        g726-40     g726-40
g723        g723sf      g723|g723sf
g719        g719        g719
23 file formats registered.
```

The syntax for MixMonitor() in the dialplan is as follows:

```
MixMonitor(filename.extension[,options[,command]])
```

The options string can contain any of the following options:

a

> If the specified filename already exists, append to it instead of overwriting it.

b

> Delay the start of the recording until the call has been bridged. Otherwise the recording will start during call setup. If you only want to record the parts of the call once both sides have been answered and are talking, use this option.

v(x)

> Adjust the volume of the audio heard by the channel that is executing `MixMonitor()`. The range is -4 to 4.

V(x)

> Adjust the volume of the audio spoken by the channel that is executing `MixMonitor()`. The range is -4 to 4.

W(x)

> Adjust the audio coming from both directions. The range is -4 to 4.

The final parameter for `MixMonitor()` is a `command`. This is a custom command that will be executed when the recording is complete. This can be useful for doing post-processing of recordings.

See Also

See Recipe 3.6 for another recipe that is related to call recording. To see the built-in documentation that Asterisk has for `MixMonitor()`, execute this command at the Asterisk CLI:

```
*CLI> core show application MixMonitor
```

3.6 Triggering Call Recording Using DTMF

Problem

You would like to give some users the ability to enable call recording by pressing a key sequence during a call.

Solution

Use the `automixmon` feature. This feature is built into Asterisk. You just have to enable it in a couple of configuration files. First, set the `automixmon` option in `features.conf` to the key sequence you would like to use for enabling or disabling recording:

```
[featuremap]
automixmon = *3
```

Now that the key sequence has been set, you must modify *extensions.conf* to allow callers to use this feature. This is done by setting the x and/or X options in the Dial() or Queue() application:

x

Give the called party the ability to toggle call recording using the automixmon feature.

X

Give the calling party the ability to toggle call recording using the automixmon feature:

```
;
; When someone calls my extension, dial my phone and give
; me the ability to enable recording on the fly.
;
exten => 7001,1,Dial(SIP/myphone,30,x)
```

Discussion

The featuremap section of *features.conf* has some other features that are configured and enabled in this same manner. You configure the key sequence for the feature in *features.conf* and then enable the feature on a per-call basis using arguments to the Dial() or Queue() applications. Here is a list of all of the features in the featuremap section:

blindxfer

DTMF triggered blind transfers. Enable it using the t and/or T options to Dial() or Queue().

atxfer

DTMF triggered attended transfers. Enable it using the t and/or T options to Dial() or Queue().

disconnect

DTMF triggered call hangup. Enable it using the h and/or H options to Dial() or Queue().

parkcall

DTMF triggered call parking. Enable it using the k and/or K options to Dial() or Queue().

automixmon

DTMF triggered call recording using the MixMonitor() application. Enable it using the x and/or X options to Dial() or Queue().

automon

DTMF triggered call recording using the Monitor() application. We recommend using the automixmon feature instead of this one unless you have a specific need for using Monitor() instead of MixMonitor(). Enable this feature by specifying the w and/or W options to Dial() or Queue().

See Also

For another recipe that uses `MixMonitor()`, see Recipe 3.5.

3.7 Making Grandma Louder

Problem

You have a particular caller who is soft spoken, and need to increase the volume of her speech.

Solution

```
[VolumeAdjustment]
exten => 100,1,Verbose(2,Incoming call from ${CALLERID(all)})
    same => n,GoSubIf($[${CALLERID(num)} = 12565551111]?VolumeAdj,1)
    same => n,Dial(SIP/0000FFFF0001,30)
    same => n,Hangup()

exten => VolumeAdj,1,Verbose(2,Adjusting volume for grandma)
    same => n,Set(VOLUME(TX)=3)
    same => n,Return()
```

Discussion

In our solution, we've used the `CALLERID()` function to match on a particular callerID number and, if matching, to execute the `VolumeAdj` extension which increases the receive volume. We're making use of a `GoSubIf()` to execute the `VolumeAdj` extension and to return when the volume adjustment is completed, and then dialing the device 0000FFFF0001. We could then create a list of callerID numbers to adjust the volume for if we wished. Because we're expecting the other side to be quieter than we would like, we are increasing the volume of the transmitted audio by specifying `VOLUME(TX)`. If we wanted to increase the volume for the audio received by that channel, we would use `VOLUME(RX)`.

> The `VOLUME()` function must be viewed from the aspect of the channel that is executing the function. Because there are two channels in every call, if the channel initiating the `Dial()` application executed the `VOLUME()` function, then the RX option adjusts their receive volume, and TX adjusts their transmit volume.

In addition to changing the volume prior to calling an application, with the p option we can permit the `VOLUME()` function to listen for DTMF to adjust the volume of the channel. To increase volume, use the * key. To lower the volume, use the # key. With our current dialplan, the DTMF can be adjusted by the caller, which may not necessarily be what we want. If we want the `VOLUME()` to be adjustable on the called channel, we need to execute a subroutine on the other channel just prior to bridging. We can do this with the U() option to the Dial() application:

```
exten => 100,1,Verbose(2,Incoming call from ${CALLERID(all)})
same => n,Dial(SIP/0000FFFF0001,30,U(VolumeAdjustment^3))
same => n,Hangup()

exten => s,1,Verbose(2,Adjusting volume for other channel)
same => n,Set(VOLUME(RX,p)=${ARG1})
same => n,Return()
```

 We could use the IF() function to control whether the subroutine is executed by the Dial() application:

```
same => n,Dial(SIP/0000FFFF0001,30,${IF($[${CALLERID(num)} = 12565551212]?
U(VolumeAdjustment^3))})
```

The nice thing about adjusting the volume with the VOLUME() function is that you can apply this not just to dialing end points, but also to adjust volume prior to sending calls to other applications which may be recording audio, such as Voicemail() or Record(). In this manner, you can adjust the volume for people individually but continue utilizing the same dialplan logic.

If you had a large list of people with different volumes you needed to adjust, then making use of the AstDB or *func_odbc* methods of looking up information in a database would be the better way to go. By making the data dynamic, you can handle tens or hundreds of different volume configurations without much more dialplan. If using func_obdc and a custom function, the dialplan may look something like the following:

```
[VolumeAdjustment]
exten => 100,1,Verbose(2,Incoming call from ${CALLERID(all)})
    same => n,Set(ARRAY(VolLevel,VolDirection)=
${ODBC_GET_VOLUME_LEVEL(${CALLERID(num)})})
    same => n,GoSubIf($[${EXISTS(${VolumeLevel})}]?VolumeAdj,1)
    same => n,Dial(SIP/0000FFFF0001,30)
    same => n,Hangup()

exten => VolumeAdj,1,Verbose(2,Adjusting volume for grandma)
    same => n,Set(VOLUME(${VolDirection})=${VolLevel})
    same => n,Return()
```

Once you start abstracting the data with things like *func_odbc* then you can even control volume based on who is being called (perhaps someone has impaired hearing and requires the volume to be adjusted) and all sorts of other situations.

Get even more for your money.

Join the O'Reilly Community, and register the O'Reilly books you own. It's free, and you'll get:

- $4.99 ebook upgrade offer
- 40% upgrade offer on O'Reilly print books
- Membership discounts on books and events
- Free lifetime updates to ebooks and videos
- Multiple ebook formats, DRM FREE
- Participation in the O'Reilly community
- Newsletters
- Account management
- 100% Satisfaction Guarantee

Signing up is easy:

1. Go to: oreilly.com/go/register
2. Create an O'Reilly login.
3. Provide your address.
4. Register your books.

Note: English-language books only

To order books online:
oreilly.com/store

For questions about products or an order:
orders@oreilly.com

To sign up to get topic-specific email announcements and/or news about upcoming books, conferences, special offers, and new technologies:
elists@oreilly.com

For technical questions about book content:
booktech@oreilly.com

To submit new book proposals to our editors:
proposals@oreilly.com

O'Reilly books are available in multiple DRM-free ebook formats. For more information:
oreilly.com/ebooks

Spreading the knowledge of innovators   oreilly.com

©2010 O'Reilly Media, Inc. O'Reilly logo is a registered trademark of O'Reilly Media, Inc. 00000

The information you need, when and where you need it.

With Safari Books Online, you can:

Access the contents of thousands of technology and business books

- Quickly search over 7000 books and certification guides
- Download whole books or chapters in PDF format, at no extra cost, to print or read on the go
- Copy and paste code
- Save up to 35% on O'Reilly print books
- **New!** Access mobile-friendly books directly from cell phones and mobile devices

Stay up-to-date on emerging topics before the books are published

- Get on-demand access to evolving manuscripts.
- Interact directly with authors of upcoming books

Explore thousands of hours of video on technology and design topics

- Learn from expert video tutorials
- Watch and replay recorded conference sessions

O'REILLY®

Milton Keynes UK
Ingram Content Group UK Ltd.
UKHW031814010823
426163UK00007B/355